TAIWAN

MAJOR WORLD NATIONS
TAIWAN

Jessie Wee

CHELSEA HOUSE PUBLISHERS
Philadelphia

Chelsea House Publishers

Copyright © 1999 by Chelsea House Publishers,
a subsidiary of Haights Cross Communications.
All rights reserved.
Printed in China

3 5 7 9 8 6 4 2

Library of Congress Cataloging-in-Publication Data

Wee, Jessie.
Taiwan / Jessie Wee.
p. cm. — (Major world nations)
Includes index.
Summary: Explores the people, history, climate, economy, and
topography of Taiwan, the island province of China that continues to
resist communism.
ISBN 0-7910-4986-8
1. Taiwan—Juvenile literature. [1. Taiwan.] I. Title.
I. Series.
DS799.W44 1998
951.24'9—dc21 98-4308
CIP
AC

ACKNOWLEDGEMENTS

The Author and Publishers are grateful to the following organizations and individuals
for permission to reproduce copyright illustrations in this book:
J. Allan Cash Photolibrary; Hutchison Photo Library; Peter Newark's Historical
Pictures; Helene Rogers; Taiwan Visitors Association, Singapore; Travel Photo
International.

CONTENTS

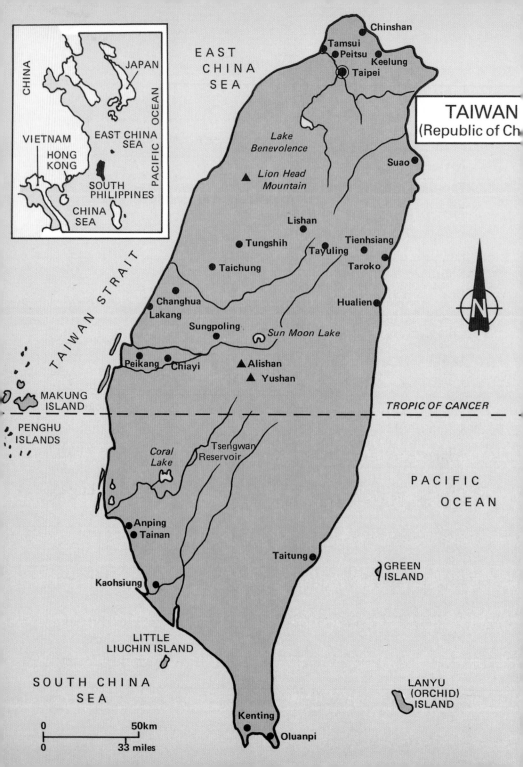

CHINA

JAPAN

VIETNAM

HONG KONG

EAST CHINA SEA

SOUTH PHILIPPINES

CHINA SEA

PACIFIC OCEAN

TAIWAN
(Republic of Ch

EAST CHINA SEA

Chinshan

Tamsui
Peitsu

Keelung

Taipei

Suao

Lake Benevolence

▲ Lion Head Mountain

Lishan

Tungshih

Tienhsiang

Tayuling

Taichung

Taroko

Changhua

Hualien

Lakang

Sungpoling

Sun Moon Lake

Peikang

Chiayi

▲ Alishan

▲ Yushan

TAIWAN STRAIT

TROPIC OF CANCER

MAKUNG ISLAND

PENGHU ISLANDS

Coral Lake

Tsengwan Reservoir

PACIFIC OCEAN

Anping
Tainan

Taitung

GREEN ISLAND

Kaohsiung

LITTLE LIUCHIN ISLAND

LANYU (ORCHID) ISLAND

SOUTH CHINA SEA

0   50km

0   33 miles

Kenting

Oluanpi

FACTS AT A GLANCE

Land and People

Official Name	Republic of Taiwan
Location	Located off the southeast coast of the China mainland; bound by the Pacific Ocean to the east
Area	13,900 square miles/ 36,000 square kilometers
Climate	Subtropical in the north/ tropical in the south
Capital	Taipei
Other Cities	Kao-hsiung, T'ai-chung, T'ai-nan, Chi-lung
Population	22 million
Population Density	1,547 persons per square mile (97.5 persons per square kilometer)
Major Rivers	Tan, Cho
Major Lakes	Coral Lake
Mountains	Central Mountain Range, T'ai-tung
Official Language	Mandarin Chinese
Other Languages	Taiwanese dialects of Chinese

7

Ethnic Groups	Han Chinese, Malayo-Polynesian aborigines
Religions	A mixture of Buddhism and Taoism is the most practiced religion
Literacy Rate	93.7 percent

Economy

Natural Resources	Petroluem, natural gas, coal
Agricultural Products	Rice, maize, soybeans, sugar cane
Industries	Manufacturing, construction, mining, pharmaceutical
Major Imports	Electronic machinery, chemical, iron, steel and motor vehicles
Major Exports	Nonelectrical machinery, electrical machinery, plastic, synthetic fibers
Major Trading Partners	Japan, the United States, Germany
Currency	New Taiwan dollar

Government

Form of Government	Multi-party form of government
Government Bodies	Taiwan National Assembly and five *yuan* (councils)
Formal Head of State	President
Head of Government	Prime Minister

HISTORY AT A GLANCE

10,000 B.C.	Prehistoric people arrive on the island, both from China and Southeast Asia.
4000 B.C.	The Lung-shan, an advanced Stone Age culture from the mainland, also flourishes on Taiwan. In general, Taiwan is unaffected by events on the mainland for many centuries.
206 B.C.	First mention of Taiwan in Chinese literature, where it is referred to as Yangchow or Yichow.
239 A.D.	First attempt to establish a Chinese claim to Taiwan, by the kingdom of Wu.
500	The Hakka, a mainland ethnic minority, are driven from China and establish themselves on Taiwan.
1400s	The name "Taiwan," meaning "terraced bay" first begins to appear in Chinese documents.
1368-1644	During the Ming Dynasty, there is increasing Chinese immigration, generally from Fujian province.
1550s	Japanese pirates raid the coasts of Taiwan, the beginning of a long-standing Japanese interest in the area.

1590	The Portuguese establish a trading post on Taiwan, which they call the "Isla Formosa," or "beautiful island."
1624	The Dutch establish a trading post, Zeelandia, on Taiwan.
1646-1658	Taiwan's first Chinese ruler, Koxinga, launches attacks on the Manchus in an attempt to restore the Ming Dynasty.
1652	A Taiwanese uprising against the Dutch is suppressed with great slaughter.
1661-1662	Koxinga expells the Dutch from the island. He devotes the rest of his reign to preserving Chinese culture.
1683	Death of Koxinga.
1684	The Manchus establish control.
1858	After the second of the opium wars, the Western powers force China to sign the Treaty of Tientsin. Among other provisions, the treaty forces China to open four Taiwanese ports to Western trade.
1886	Taiwan declared province of China, ending status as a prefecture of mainland Fujian province.
1894-1895	Japan wins the Sino-Japanese War. As part of the peace settlement, Japan acquires Taiwan, which it will keep until 1945. During this period Taiwan is generally known in the West by its Portuguese name "Formosa."
1941	The extension of World War II into Asia gives China allies against the Japanese. Britain and the United States support China's Nationalists under Chiang Kaishek.

1943	At the Cairo Conference, Winston Churchill and Franklin Roosevelt promise Chiang Kaishek the return of Japanese-controlled Manchuria and Taiwan to China after the war.
1945	After Japan surrenders, Chinese sovereignty over Taiwan is restored (October 30).
1947-1949	The end of World War II leads to a civil war in China between the Communists and Nationalists, who had united against the Japanese. The Communists eventually occupy all of Mainland China. Chiang Kaishek sets up a government-in-exile on Taiwan, which calls itself the Republic of China. The Communist mainland regime calls itself the Peoples Republic of China.
1955	Taiwan and the U.S. sign the Sino-American Mutual Defense Treaty.
1965	U.S. economic aid ceases. Over the next decade, political liberties remain at a low level, but the developing economy gradually opens up the society.
1971	The United Nations seats mainland China as the sole representative of the Chinese people, and Taiwan is expelled.
1975	Chiang Kaishek dies on April 5.
1978	The United States officially recognizes mainland China. As a part of the accord with the mainland, the U.S. ceases official diplomatic relations with Taiwan.
1990	National Assembly elects Lee Teng-hui to full term as president, the first native-born Taiwanese to

hold the office. Lee begins instituting political reforms.

1993   First official government talks between China and Taiwan. Political and diplomatic talks stall, but economic relations grow dramatically.

1996   First popular election for president of Taiwan; previous presidents had been elected by the National Assembly. Lee Teng-hui wins. Mainland Chinese military forces conduct threatening maneuvers and U.S. naval forces also enter the area. The situation returns to normal after the election.

1

A Colorful Past

To the Chinese, dragons are a symbol of strength and goodness. In former times, the Chinese Emperor was closely associated with the Dragon God and his throne was known as the Dragon Throne.

Traditionally connected with water, dragons were believed to be found in pools, rivers or the sea. According to a legend, these noble creatures were responsible for the creation of Taiwan.

Many years ago, when the earth was still young (so the story goes), dragons played off the coast of China. Their huge, thrashing tails stirred up the ocean bottom and out of the depths rose the island of Taiwan. The dragons used the island as a place to rest and sleep, and are believed to be there still. The aborigines (descendants of the original inhabitants of Taiwan) say it is the snores of the dragons that can be heard in the fissures and caves found on the island, and it is dragons' breath that fills the air with sulphur fumes in its northernmost parts.

Geologists believe that volcanic action pushed the island up from under the sea. Evidence of this volcanic activity, which ended

A mural painting of a dragon by Taiwanese schoolchildren. Legend has it that dragons were responsible for the creation of Taiwan.

millions of years ago, can be found in the bubbling pools of sulphurous water, the hissing steam vents and the coral rock formations on the island. In Taiwan today, earthquakes still occur throughout the year, but they are so minor they are seldom felt.

Ancient records show that the Chinese were aware of Taiwan, which was then known by a different name, as long ago as two hundred years before the birth of Christ. Attempts to explore the island may have been made during these early times.

In the thirteenth century, Taiwan became a protectorate of the Chinese Empire when Genghis Khan, the Mongol conqueror, made himself the emperor of China. However, the island was not given the name of Taiwan until 1430 A.D. when the famous Admiral Cheng Ho, from the Ming court, reported his "discovery" of the island to the Ming emperor.

14

Tribes of aborigines have lived in Taiwan since prehistoric times. They are believed to have come from different parts of Asia. Many were a small, brown-skinned people who were fierce and warlike.

By the time the Chinese settlers came to the island, some aboriginal groups had already settled down to an agricultural life in the fertile plains of central and southwestern Taiwan. Others roamed the mountainous regions, fighting among themselves and carrying out their practice of head-hunting.

No one knows exactly when the Chinese first began to settle in Taiwan. The first immigrants are believed to have been the Hakkas, a minority group persecuted in China since ancient times. The Hakkas, driven from their homes in Honan province, fled to the Fukien and Kwangtung coasts in the southeastern part of China. Their fishing and trading activities brought them to the Penghu Islands, off the mid-western coast of Taiwan, and then to Taiwan itself.

The Hakkas drove the aboriginal tribes from the fertile plains into the mountains. They, in turn, were pushed further inland by later immigrants from the Fukien province of China.

Although Taiwan had first been sighted by the Portuguese, the first Westerners to set foot on the island were the Dutch. In 1624, they set up their capital, Fort Zeelandia, on the southwest coast—on the site which is now Anping. The Dutch opened trading-posts, introduced Christianity to the people, and encouraged the planting of sugarcane and the extraction of camphor from the camphor trees growing on the island.

The Spaniards, wanting to establish new trade routes in the Far

East, set up small communities in northern Taiwan two years later. But the Dutch had no intention of letting another European power share in their new colony and its trade. They drove out the Spaniards in 1642.

In this year, too, the Manchu conquest of China began. It was, as we shall soon see, to change the future history of Taiwan.

From Manchuria in the north, the Manchus swept down into China to defeat the armies of the Ming emperor. China's northern capital, Peking (now known as Beijing), fell to the Manchus in 1644. Unable to stem the tide of Manchu victories, a group of Ming loyalists, under the leadership of Cheng Chengkung, retreated to Taiwan in 1661. They seized Fort Zeelandia and expelled the Dutch from the island. Cheng Cheng-kung, who was also known as Kuo Hsing-yeh ("Lord of the Imperial Surname"), is better known to the West as Koxinga. He gave Taiwan its first Chinese government. Many traditional customs and lifestyles of the Chinese were established on the island, for among Koxinga's followers were more than "one thousand carefully chosen scholars, artists, monks and masters of every branch of Chinese culture."

From Taiwan, Koxinga—and later his son and grandson—continued to defy Peking long after the Manchus had gained control of the whole of mainland China.

The Ming loyalists never realized their dream of retaking mainland China from the Manchus, but they opened up Taiwan for settlement by large groups of Chinese, and made the island a flourishing center for trade. They ruled Taiwan for twenty-three years, until 1684, when the island finally fell to Manchu forces from the mainland.

16

A Chinese woman and her child in Taipei, the capital of Taiwan. It was in the seventeenth century that the Chinese first began to settle in Taiwan in large numbers and today the vast majority of Taiwanese are of Chinese descent.

The Manchus set up the Ching dynasty which was to rule China for the next 268 years under a succession of ten emperors. The mighty Chinese empire stretched from Burma to Korea with its northern capital at Peking and its southern capital at Nanking (Nanjing).

The beginning of the nineteenth century saw Manchu power slowly crumbling because of weak and corrupt rulers. As a result, powerful Western countries succeeded in forcing China to open its important ports to them. Four of Taiwan's ports were also opened to Western traders and missionaries.

Many British and American firms were set up in Taiwan. Trade increased. In quarrels with China during this period, the Japanese briefly occupied southern Taiwan and the French its northernmost parts, as well as the Penghu Islands.

In 1895, Japan invaded Korea. The Chinese army, which had come to the aid of Korea, was defeated and, under one of the clauses of the Treaty of Shimonoseki, China ceded Taiwan and the Penghu Islands to Japan. For the next fifty years, Japan ruled Taiwan with an iron hand.

The Japanese were not interested in colonizing Taiwan. They regarded the island merely as a source of essential foodstuffs such as rice and sugar. The island was also a convenient spot for an industrial, and later, a military base for Japan's conquest of Southeast Asia. The Japanese rulers forced the people of Taiwan to sever their Chinese cultural roots, learn to speak Japanese and adopt Japanese names. Revolts in various parts of the island were all successfully put down by the Japanese army.

In spite of their harsh rule, the Japanese were respected for bringing law and order to the island. They introduced better methods of agriculture and built modern schools and hospitals. They built railways and a fine system of paved roads to link the different parts of the island. Textile mills, paper mills, fertilizer plants, oil refineries and other industries were also started.

Meanwhile, in China, things were going from bad to worse. Finally, in 1911, the Chinese people rebelled, causing the downfall of the Manchu Ching dynasty and, with it, the end of five thousand years of imperial rule in China.

China was declared a republic on January 1, 1912, and Dr. Sun Yat-sen, the leader of the Nationalist movement, was made its first provisional president.

The Nationalists, under the Kuomintang ("People's Party"), began to introduce new reforms. To guide them, they used the three principles which Dr. Sun Yat-sen considered important as the framework for a sensible government. These three principles are: Nationalism (or independence), Democracy (or government by the people), and Socialism in which the State is responsible for the education, health and welfare of its people.

The Nationalists were particularly strong in the south of China where they had made Nanking their seat of government. The rest of the country was under the control of local military leaders or "warlords" who, disagreeing with the way the new Republic of China was being ruled, lost no time in seizing power for themselves.

When Dr. Sun Yat-sen died in 1925, it was left to his successor, Chiang Kai-shek, to bring about a peaceful and united China. In 1928, Chiang Kai-shek led the Nationalist army northwards into Peking after a successful campaign against the warlords. China was once more united, but not for long, for the Nationalists soon found themselves at war against the Communists and the invading Japanese.

Disagreement within the Nationalist party had caused a split between the Conservative and Communist sections. The Conservatives, under Chiang Kai-shek, succeeded in driving the

Dr. Sun Yat-sen.

Communists into exile. The Communist movement, however, attracted the landless peasants in the central provinces.

In 1931, Japanese forces attacked the Chinese army in Manchuria, and in 1933 pushed further south into Chinese territory. During a brief period of peace between China and Japan, fighting broke out again between the Conservative Nationalists and the Communists. Both factions, however, agreed to unite to fight the Japanese who attacked China again in 1937.

20

In 1941, the Japanese attacked Pearl Harbor, a U.S. naval base in Hawaii. This led to the United States of America entering the Second World War. At the same time, they became Chiang Kai-shek's ally against Japan. When the Second World War ended in 1945, with the defeat of Japan, Taiwan and the Penghu Islands were returned to China. But there was to be no peace on the mainland, for civil war broke out between the Nationalists and the Communists. It dragged on for four years until 1949 when the Communists finally took over mainland China.

Chiang Kai-shek and his followers retreated to Taiwan, making

Chiang Kai-shek. When the Communists took over mainland China in 1949, Chiang Kai-shek and his followers retreated to Taiwan and set up a government based on Dr. Sun Yat-sen's "Three Principles of the People."

Taipei the seat of government of the Republic of China. There are striking similarities between Chiang Kai-shek's retreat to Taiwan and that of Koxinga more than three hundred years ago.

Both leaders fought to preserve traditional order in mainland China. When this was not possible, they held out in Taiwan in defiance of their enemies on the mainland. During their exile in Taiwan, both leaders kept alive China's ancient and cherished culture. Among their followers were some of the best Chinese scholars, artists and craftsmen.

Chiang Kai-shek governed Taiwan, now known as the Republic of Taiwan, according to Dr. Sun Yat-sen's "Three Principles of the People." He maintained law and order, introduced successful land reforms, improved the education system, and gave the people a local, democratic government while keeping state affairs in the hands of the Nationalist government.

Assisted in the early years by its powerful ally, the United States, Taiwan continued to modernize, and to make great progress in its industry and economy.

Chiang Kai-shek died in 1975. In 1978, his son, Chiang Ching-kuo, became the republic's new president. He was reelected as president six years later, in 1984. Chiang died in 1988, and was succeeded by Lee Teng-hui. Lee was the island's first Taiwanese-born president. Through the early 1990s, Taiwan continued to thrive as it maintained foreign trade.

First official government contacts between Taiwan and China took place in 1993, resulting in the signing of some agreements.

A statue of Chiang Kai-shek, the first president of Taiwan.

However, a second set of talks scheduled for 1995 was postponed indefinitely by China.

In 1996, Lee Teng-hui was reelected in the first popular election for the presidency of Taiwan. Just before the election, China initiated military manuevers aimed to scare Taiwan's voters. The launching of missiles near the seaports of Keelung and Kaohsuing, resulted in the sending of an American naval aircraft carrier into the Taiwan Strait.

2

The Beautiful Island

For hundreds of years, Taiwan was known to the West as Formosa, a name given to it by sixteenth-century Portuguese sailors who, on catching sight of the island from their ship, shouted in admiration, *Ilha Formosa! Ilha Formosa!* ("Beautiful island! Beautiful island!")

Taiwan means "terraced bay" in Chinese. Officially, it is known as "the island province of the Republic of China."

The island of Taiwan is indeed beautiful with its forest-fringed mountains, lush valleys, terraced tablelands, fertile plains, sandy beaches and rocky coastlines. Seen from an airplane, it resembles a long, slender leaf with its tip pointing towards the islands of Japan in the northeast, and its stem to the islands of the Philippines in the south.

Taiwan is 250 miles (390 kilometers) long and 90 miles (140 kilometers) wide at its broadest point. To its west lies the Taiwan (Formosa) Strait—a 100-mile (161-kilometer) stretch of water that separates the island from the southeast coast of mainland China. To the north, its shores are washed by the East China Sea, to the east

Fishing at Yehliu in northern Taiwan.

by the Pacific Ocean and to the south and southwest by the South China Sea.

The republic is made up of one large island and about eighty small offshore islands. Of these, sixty-four are in the Penghu group which lies off Taiwan's western coast.

Another twenty-one nearby islands are in the Taiwan group. Two other small island groups, Kinmen (Quemoy) and Matsu, are just off the southeast coast of mainland China. About 1.4 miles (less than 2.3 kilometers) from some islands held by Communist China, Kinmen was in the past frequently shelled by the Communists–especially between 1958 and 1960. From 1961,

Yushan, or Jade Mountain, in the Central Mountain Range. At 13,113 feet (3,997 meters), this is the highest peak in Northeast Asia.

both sides stopped firing explosive shells into each other's territory and took to firing propaganda materials instead. Communist materials included quotations from Mao Tse-tung, the first Chairman of Communist China, while Nationalist materials included pictures of the happy, contented people of Taiwan.

Taiwan's total land area is 13,900 square miles (nearly 36,000 square kilometers). This is about the size of Holland. The Tropic of Cancer runs across Taiwan just slightly south of its midpoint.

Hills and mountains cover almost two-thirds of Taiwan.

Running from north to south, like the backbone of some sleeping dragon, is the Central Mountain Range. It has many peaks which are more than 10,000 feet (3,000 meters) above sea level. The highest is Yushan or Jade Mountain (Mount Morrison), in central Taiwan. It soars to a height of 13,113 feet (3,997 meters), making it the highest peak in Northeast Asia.

The eastern slope of the Central Mountain Range is steep and rugged. In some areas it drops sharply into the Pacific Ocean. Some of the world's highest sea cliffs can be found here. The western slope descends more gradually into a broad, coastal plain.

All Taiwan's rivers rise in the Central Mountain Range. They are short and swift, and most of them flow westward to the coastal plains. Boats, unfortunately, cannot ply up and down all the rivers. This is because the rivers have no steady currents and are often shallow during the dry season and flooded during the wet. But most of the rivers are useful as they bring water and fertile alluvial soil to the plains. A number provide hydroelectric power.

The island's position, its geographical features and the warm Kuroshio (Japan) Current that flows around it are responsible for Taiwan's mild climate. This climate can be described as subtropical in the north and the mountainous regions, and tropical in the south.

There are two distinct seasons—a hot season from May to October, and a cold season from November to March. The hot season (summer) is long and humid with average temperatures

A dam near the East-West Cross-Island Highway on the Liwu River, used as a source of hydroelectric power.

that vary, in places, from 80 to 90 degrees Fahrenheit (27 to 32 degrees Celsius). The cold season (winter) in the north is short and mild, with temperatures ranging between 50 and 60 degrees Fahrenheit (10 and 15 degrees Celsius). Although winter is short, the months of January and February are often cold enough to bring snow to the mountain tops.

There is plenty of rainfall throughout the island, with the upland areas receiving more rain than the lowlands, and the eastern part of the island receiving more than the western part. The winter monsoons (seasonal winds) bring rain to northern Taiwan while the summer monsoons bring rain to southern Taiwan. The average rainfall per year is between 40 and 60 inches (1,000 and 1,500 millimeters) on the west coast. In some mountainous area, it is more than 200 inches (5,000 millimeters).

Taiwan lies in the Pacific typhoon belt. The Chinese call the typhoons or violent hurricanes of the China seas *chu-feng* which means "wind from all quarters." These typhoons rise in the Indonesian islands in the south. They storm northwards through the Philippines towards Japan, via Taiwan. Most of the storms miss Taiwan but sometimes a typhoon crashes into the island with

A roadside stall selling fruit known locally as "Lotus in the Fog." Taiwan is situated in the Tropics and is thus able to grow many unusual tropical fruits.

wind speeds of 100 miles (160 kilometers) per hour or more. Such violent storms can capsize huge ships, blow down trees and dwellings, and even flood low-lying towns and cities. The dreaded typhoons occur from May to October, with July, August and September being the worst months.

Taiwan's flag consists of a white sun in a blue sky over a crimson background. Its national flower is the plum blossom which blooms in winter. Plum blossoms are often featured in Chinese paintings for they are a symbol of strength, courage and perseverance. The petals of the flower represent the five *yuan* or councils of the government, while its three buds symbolize the "Three Principles of the People."

There are many political parties in Taiwan today. All are strongly against Communism. The oldest party, the Kuomintang,

Clearing fallen trees and rocks from Taiwan's East Coast Highway after a typhoon—typhoons occur from May to October.

The entrance to the Taiwan National Assembly.

has been in power since 1949 when it made Taipei the seat of the Nationalist government.

The Taiwan constitution is based on the principles "of the people, by the people, for the people." Until 1996, the people elected their representatives to the Taiwan National Assembly which, in turn, elected the president or head of state of the republic. But the first popular election for president of Taiwan took place in 1996.

The president is the supreme leader of the nation. Under him are five independent *yuan* or councils of the government. The *yuans* include the Executive *Yuan* which is a council resembling the President's cabinet in the United States; the Legislative *Yuan*

31

Chung Hsing Village—the seat of provincial government in Taiwan.

which makes the laws; the Judicial *Yuan* which includes the courts and the Council of Grand Justices; the Examination *Yuan* that supervises government personnel and examinations; and the Control Yuan that has powers of censure, impeachment and audit. The head of government is the premier or prime minister. There are also numerous county and city councils that are elected by the people.

3

The People of Taiwan

Taiwan today has a population of over twenty-two million people. Most are Chinese. They live mainly on the western coastal plains where the major cities and towns are situated.

The aborigines, descendants of the first or original people on the island, number more than 375,000—slightly over one percent of the total population. They live mostly in the remote valleys

Girls of the Ami tribe, the largest group of aborigines in Taiwan.

and on the slopes of the Central Mountain Range. Today, only nine main tribes remain in numbers large enough to keep their customs and cultures alive. The others have been absorbed into the modern, Chinese society of Taiwan.

The Ami form the largest group among the aborigines with more than 150,000 members. They live in the mountains and valleys near Hualien on the east coast. The Atayals, the second largest tribe in Taiwan, live in the mountainous region of Wulai, to the south of Taipei.

In the foothills and mountainous regions of central and southern Taiwan live other aboriginal tribes such as the Paiwan, Puyuma, Rukai and Bunun. Each has its own language, customs and culture. Among them, the Paiwan are master woodcarvers. They carve totems and other works of art. They are also expert weavers. Their traditional methods and ancient designs form a part of their heritage to this very day. The only seafaring tribe of aborigines are the Yami. They live in Lanyu or Orchid Island off the southeast coast of Taiwan.

Apart from the aborigines, the rest of the Taiwanese are largely descendants of the Chinese who migrated to Taiwan from all parts of mainland China. They are made up of two distinct groups—those native to Taiwan, and the "mainlanders" or recent arrivals.

The native Taiwanese are the descendants of Chinese immigrants from the Fukien and Kwangtung provinces of mainland China. Large numbers of these immigrants came and settled in Taiwan between the sixteenth and nineteenth

An old Chinese woman in Taipei—most Taiwanese are of Chinese descent but many of them are able to speak Japanese.

centuries. Smaller numbers of them may have come over much earlier.

The different groups from these provinces had different habits and spoke different dialects. Since more immigrants came from the Fukien province, their Amoy dialect is widely spoken by the Taiwanese today, especially in the rural areas.

During the Japanese occupation of Taiwan (1895-1945), the Taiwanese were forced to acquire the language, education and customs of the Japanese. It is not surprising, therefore, to find that many of the older Taiwanese are able to speak Japanese.

The "mainlanders," or recent arrivals, are the Chinese who fled to Taiwan during the Communist takeover of mainland China in

1949. They live mainly in the Taipei area and other urban centers and make up about eighteen percent of Taiwan's population. They come from all parts of mainland China and speak Mandarin, which is today the official language of the people of Taiwan. All these different groups of Taiwanese have, over the years, gradually come together through the increased use of Mandarin, a progressive education program, and intermarriage.

It is interesting to note that although the Chinese speak many different dialects, and may not understand one another's spoken language, they have only *one* written language. This means that no matter which part of China the people come from or what dialect

A colorful anti-crime poster. There is no alphabet in the Chinese language—instead, each character represents an object or an idea.

they speak, they can *read* Chinese once they are taught the written form.

This written form is based on "idea-pictures" or ideograms as they are called. Each character or word, with its single sound, is used to represent an object or an idea, as there is no alphabet in the Chinese language.

China has a civilization which has lasted longer than that of any other country on earth. At least five thousand years old, its culture has produced a written language that is still in use today with only some minor changes. There are as many as fifty thousand characters in the Chinese language. Already well-developed almost fifteen hundred years before the birth of Christ, this written language has been an important force in uniting the Chinese people. Chinese is the oldest and most commonly used language in the world today. About three times more people speak Chinese than English, the next most commonly used language.

Unlike Communist China, Taiwan still uses the standard written form of the Chinese language. To help foreigners, the Taiwanese spell out Chinese words in the Roman alphabet (romanized script) according to the most common English pronunciation. (Communist China has, since 1958, adopted a simplified form of Chinese characters. Using the Roman alphabet, it has also worked out its own system of teaching the way each character is pronounced. This is known as the Pinyin system.)

Education has always been held in high regard by the Chinese.

Confucius (551-479 b.c.). China's greatest philosopher, whose teachings are still adhered to in Taiwan.

Everyone who could made sure that their sons learned to read and write. From the earliest times, the gentleman-scholar who devoted his life to the pursuit of knowledge was highly honored and respected.

Confucius (551-479 B.C.) was China's greatest teacher. He said, "By nature men are pretty much alike; it is learning and practice that set them apart." Confucius believed in peace and social harmony, the importance of family and friends, and the virtues of self-discipline and kindness towards others. He travelled all over the country to teach his ideas and was responsible for laying the foundation of the educational system in China. For more than two thousand years, Chinese scholars had to memorize *The Analects*–the most famous collection of the teachings and sayings of Confucius. His teachings are still adhered to, especially in Taiwan.

Confucianism is often mistaken for a religion which it is not. It is actually a practical way of maintaining order and ruling a huge country like China.

Today, in Taiwan, free education is available to all children for nine years. Six years are spent in the primary or elementary school and three in the junior (middle) school. Then, there are the senior (secondary) schools where general or vocational education is available. Higher education is available in the junior colleges, universities and research institutes.

Education in Taiwan places great importance not only on scientific knowledge but also on cultural traditions, and on the values and conduct held in high regard by the Chinese people. Besides reading, writing, mathematics, science and other subjects, Taiwanese children learn about Confucius and other great men in Chinese history, the value of work and how important it is for citizens to be of use to the community. The language of instruction in Taiwanese schools is Mandarin. English is a compulsory subject in middle and senior school, but not everyone is fluent in the language.

Schoolchildren take an active part in sports. There are track and field events to prepare the young for future Olympics. There are also other sports such as tennis, swimming, mountain climbing, hiking and table-tennis. National sports include baseball, soccer and basketball. Young and old alike have

The "flag ceremony" (to salute the Taiwanese flag) carried out by children in a school playground in Taipei.

taken to the sports popular in the West, including golf.

Many sports and competitions were introduced to Taiwan from the West at the end of the nineteenth century. But the people had not given up their love of traditional Chinese sports.

Chinese sports, which consist mainly of the martial arts— hand-to-hand combat—is very different from that of the West. Martial art movements are as graceful and fluid, yet as sudden and swift, as those of fighting creatures in the wild.

The Chinese developed their various styles of martial art long before the young boys of Sparta, in ancient Greece, were taught

40

to throw the javelin. Martial art secrets and skills were handed down from master to disciple for generations. Only carefully selected disciples were taught the skills, more as a spiritual development than a show of force. These skills were brought to Taiwan by successive groups of Chinese immigrants.

The Chinese martial art known as *kung-fu* in the West is practiced in Taiwan as variations of *tai chi chuan* or shadow boxing. The old and the young can be seen practicing the graceful movements of *tai chi chuan* in the city parks every morning. It is a very good form of exercise. It keeps the body supple, the mind relaxed, and the breathing controlled.

As an important center for trade in the earlier centuries, the island of Taiwan attracted not only the peoples from the surrounding regions but also those from the West. So it is no

Early morning exercises in a park in Taipei. The traditional Chinese sports are still widely practised in Taiwan today.

surprise to find that religions from many parts of the world have found their way to the island.

The aborigines brought their many different forms of worship and the Chinese their beliefs in Buddhism, Taoism and Confucianism. The Dutch introduced Protestant Christianity when they came to the island. The Spaniards brought their own Roman Catholic religion; and the Japanese introduced Shintoism into Taiwan when they ruled the island.

Islam was brought into Taiwan by Chinese Muslims. Large numbers of Arab traders, Persians and Turks, who settled in the northern and western provinces of China after A.D. 651, were mostly Muslims. Through the centuries, they adopted Chinese names and ways. By the time of the Ming dynasty, they had been completely absorbed into Chinese civilization. Persecuted by the Manchus and later the Communists, many fled to Taiwan, bringing their religion with them.

The religion practiced by the majority of the Taiwanese today is a mixture of Buddhism and Taoism. Buddhism arose from the teachings of the Buddha in India about five hundred years before the birth of Christ. It was brought to China by traders.

Taoism developed in China itself. It believed in following the Tao or "Way" in which Nature works. Later, it adopted many features from the Buddhist teachings that had been brought into China. The founder of Taoism was Lao-tze who lived at about the same time as Confucius.

Many Taiwanese include the worship of ancestors, folk heroes, and deities (gods and goddesses), as well as superstitious beliefs and the teachings of Confucius, in their religion. Many simply call this interesting mixture "Buddhism"—a term which often confuses the outsider. But this form of worship is similar to that practiced by Chinese communities in Hong Kong, Macao, Singapore and other parts of Southeast Asia.

There are more than five thousand temples in Taiwan today. They range from small shrines containing a few images or tablets, to large temples with several main halls. Many of these temples, built originally during the eighteenth and nineteenth centuries by Chinese craftsmen, have been renovated over the years. Although there are separate Buddhist, Taoist and Confucian temples, the people usually combine the practices of each with their own forms of worship.

Buddhist temples, as a rule, contain few images, and these are usually of the Buddha. Confucian temples contain no images at all. The Taoists have hundreds of gods and it is usually in their temples that the greatest number of images of gods, spiritual leaders and folk heroes such as Koxinga and Confucius can be found.

Chinese temples in Taiwan and overseas are easily recognized by the downward and upward slopes of their roofs. Images on the roofs include the dragon and the phoenix or legendary heroes and deities. Each temple is usually named after the chief god or goddess on the main altar.

During religious festivals, crowds of worshippers go to the temples to pray. Joss-sticks are lit and incense burnt. Offerings such as food and drink, oil and paper money are made. Offerings

Lighting joss sticks in a Taiwanese Buddhist temple.

made to the gods and to ancestors are, however, kept separate.

In ancient times, ancestral tablets were kept in family homes. Nowadays, most ancestral tablets are placed on special altars in the temples.

Offerings are an important part of ancestral worship because the Chinese believe that the spirit world is similar to the human world. They believe that if they do not care for the needs of deceased members of the family in the spirit world, they will be letting loose "hungry ghosts" on earth during the seventh lunar month when the ghosts return for a visit. Lavish feasts and street

44

operas are usually held during this month to appease any hungry or neglected ghost.

Every year, on April 5, the Taiwanese go to the ancestral burial grounds to sweep the tombs, place fresh flowers on the graves and make their food offerings. This festival is known as *Ching Ming* (Tomb-Sweeping Day) in Taiwan. It is a public holiday and coincides with the anniversary of Chiang Kai-shek's death in 1975.

Other religious festivals in Taiwan include the birthdays of the goddesses Kuan Yin and Matsu, the god Cheng Huang and the sage Confucius. Kuan Yin, the Goddess of Mercy, is one of the most popular Buddhist deities in Taiwan. Her birthday is usually celebrated in late March.

In late April, many Taiwanese congregate at the temples devoted to Matsu, the Goddess of the Sea. Prayers and offerings are made to Matsu. She is the patron saint of Taiwan and the guardian deity of the island's fishermen.

The Chinese believe that it is the goddess Matsu who guides them safely on their journeys across the sea. Matsu was born in A.D.960, on an island off the mainland of China. She was a saintly child who, at the age of sixteen, reached out through a dream to save her two brothers from a storm at sea. This and other miracles have earned her the title of "Goddess of the Sea."

The birthday of Cheng Huang, the City God, is celebrated in mid-June with colorful processions, lion and dragon dances, and lavish feasts. Since early times, it has been believed that city gods protect inhabitants from their enemies as well as from

The anniversary of the birth of Confucius is observed in the Republic of China as Teacher's Day. The anniversary, marked by age-old rites at Confucian temples, is a national holiday.

natural disasters such as floods and fires. They also advise the Lord of Heaven and the King of Hell about the rewards and punishments to be given to each resident when he dies.

The birthday of Confucius is an official public holiday. Confucius' Birthday (or Teacher's Day, as it is called in Taiwan) is celebrated on September 28. On that day, traditional ceremonies are held in Confucian temples throughout the island. Performers play classical Chinese music on ancient musical instruments, do ritual dances dressed in formal court attire or carry out other Confucian rites which are as ancient as the sage himself.

46

The Taiwanese also celebrate the Lunar New Year. This is the biggest celebration of the year, not only in Taiwan, but in all Chinese communities throughout the world. In addition, there are other public holidays to celebrate the birthdays of Dr. Sun Yat-sen and Chiang Kai-shek, as well as the Japanese restoration of Taiwan to the Republic of China, and other major historical events.

The most important holiday in Taiwan is the Double-Ten National Day. "Double-Ten" refers to the tenth day of the tenth month, or October 10, on which the celebration is held. On this day in 1911, the Chinese people brought about the downfall of the Manchu Ching dynasty. The "Double-Ten" is celebrated with a military parade, displays of folk-dancing, martial arts and

No festival in Taiwan is complete without a colorful dragon dance such as this, or a traditional and equally colorful lion dance.

other cultural activities. There are also patriotic speeches by government leaders. The celebration is held in the huge plaza in front of the Presidential Mansion in Taipei.

In Taiwan, festivals and other celebrations such as weddings and birthdays are never complete without a feast. These feasts or dinners are usually elaborate affairs to which large numbers of relatives and friends have been invited. Groups of guests sit at round tables, and the different courses of food are carefully chosen to balance and complement each other. Chinese tea, rice wine or strong drinks are usually served during the meal. Such occasions are always noisy, festive affairs with toasts being drunk to the guest of honor, to friends, good food, good fortune and anything else which springs to mind.

Traditional foods characteristic of every province of mainland China can be found in Taiwan. These include the barbecued meats of Mongolia, the roasted duck and steamed breads of Peking, the sweet and salty dishes of Shanghai, the vinegared and pickled food of Hangchow, the seafood of Foochow, and the dishes of Szechuan with their garlic, peppers, ginger and pungent sauces. Restaurants serving a wide variety of Chinese food can be found in Taipei which has, as a result, gained the reputation of being one of the world's greatest Chinese culinary centers.

The Taiwanese use chopsticks whether they eat at home, at simple roadside stalls, in eating-houses or the more lavish restaurants. For those eating alone, or those in need of a quick meal, there is the Chinese version of "fast-food." This consists of

A restaurant in Taipei—like Chinese people everywhere, the Taiwanese eat with chopsticks.

a bowl of rice or noodles with a choice of meat, vegetables and tasty sauce.

Besides being well-known for their food, the Chinese are equally noted for their arts and crafts. It is in Taiwan, more than in other overseas Chinese communities, that China's cultural heritage is being carefully preserved.

The Taiwanese have been carrying on the traditional arts and crafts of their ancestors since their arrival in Taiwan hundreds of

49

years ago. There is Chinese music played on musical instruments such as the Chinese flute, lute, fiddle and zither.

There is also Chinese painting, and calligraphy in which handwriting is developed as an art. They go hand-in-hand, for Chinese paintings usually have writing on them—either a poem or a description of the scene painted. Even the brushes used by artists and writers, and the way Chinese ink and paint are put on to paper or silk, have remained the same over the centuries.

Dances include court, aboriginal and Chinese folk dances. There are also the Chinese and Taiwanese operas. Taiwanese opera is popular with the public. It is usually performed outdoors on elevated stages. It copies the traditional Chinese opera but uses the Fukien dialect instead of Mandarin.

In both Chinese and Taiwanese operas, performers sing, dance, mime and do acrobatics while acting out an historical event or legend. Very few stage props are used but the costumes and headgear are colorful, elaborate and glittering.

Facial make-up is heavy and the colors used give each character a distinct identity. Good, honest characters have their faces painted red or black. Cunning, crafty characters have white faces, while green is the color traditionally used for ghosts and evil spirits.

Modern drama, too, has developed, and can be heard over the radio or seen on either stage or screen. In the making of Chinese motion pictures today, Taiwan's expertise rivals that of the film industry in Hong Kong. Taiwanese film-goers can see their

**A Taiwanese opera, with typically elaborate costumes and make-up.
These operas are similar to traditional Chinese performances.**

favorite, locally produced *kung fu* movies, as can film viewers in
other parts of Asia, the United States of America and Europe.

Beautiful Chinese handicrafts are in great demand overseas.
Wood and jade carvings are among Taiwan's most popular
export items. So are its fabrics of silk and brocade, its marble and
ceramic products, its bamboo ornaments and fans, and its bronze
and lacquerware.

From the earliest times, Chinese silk-weavers, potters, carvers
and other craftsmen have produced very fine work. Pottery was
one of the finest crafts of ancient China. The finest of all Chinese

pottery is porcelain, or "china" as it is often called in the western world.

The Chinese are also excellent carvers. Their favorite stone for carving is jade, which comes in shades of green, white and cream, and is carved into beautiful ornaments such as figurines and jewelry. The Chinese love jade, which they believe has magical properties.

Many ancient skills have been passed down to the Taiwanese of today. And, in the free atmosphere of Taiwan, many young and promising artists are making great progress in developing new ways of pursuing the traditional arts and crafts of China.

4

Northern Taiwan

For a closer look at Taiwan, let us start in the north–in Taipei, the gateway to the Republic of China, or Free China as it is often called.

Almost all international flights to and from Taiwan go through the Chiang Kai-shek International Airport, twenty-five miles (forty kilometers) southwest of Taipei. Outside the airport, cars, taxis and buses wait to take visitors along the road that soon joins the Chungshan (Sun Yat-sen) Expressway. This

Chiang Kai-shek International Airport, southwest of Taipei.

Taipei, the capital and largest city in Taiwan.

modern expressway runs along the west coast, linking the port
of Keelung in the northeast to the port of Kaohsiung in the
south.

About forty minutes after leaving the airport, visitors find
themselves in Taipei, the capital and largest city in Taiwan. More
than 2.5 million people live there, making it one of the most
densely populated cities in the world.

Taipei was founded in the eighteenth century by immigrants
from the Fukien province of mainland China. It grew rapidly after
1885 when it replaced Tainan as the capital of the island. *Pei*
means "north" and *nan* means "south." So we know Taipei is in
north Taiwan and Tainan in south Taiwan.

Taipei, surrounded by mountains, was built on the banks of the
Tanshui River which flows northwards through the Taipei basin.

54

This rich, fertile area is the northern portion of the west coast alluvial plains.

In the early years, Taipei was a quiet town with rice fields and mud flats and few paved roads. Today, it is the center of a well-developed urban region. It is a major industrial area with factories that produce textiles, electronic parts, machinery, motorcycles, handicrafts and various other goods.

This very modern, twentieth-century city of Asia is the administrative, cultural and nerve center of Taiwan. Roads and railway lines connect the city with all parts of the island.

Besides the ancient temples, handicraft shops and Chinese restaurants, a place no tourist ever misses when in Taipei, is the National Palace Museum whose eight great halls exhibit the world's greatest collection of Oriental art.

In 1933, priceless Chinese art treasures were carefully packed into crates and taken out of Peking, the Forbidden City in mainland China—now known as Beijing. For sixteen years, the crates were transported from one remote area of China to another to prevent them falling into the hands of the Japanese invaders. Later, in 1948, on the eve of the Communist victory, about five thousand crates of what was believed to be the best of the Chinese imperial art treasures were shipped out of mainland China to Taiwan.

Only a fraction of this collection is exhibited today in the National Palace Museum built on a mountainside beyond the northern end of the city of Taipei. The rest is safely stored, in crates, in bomb-proof concrete tunnels built into the mountain.

An ancient Chinese woodcarving on display in Taipei's National Palace Museum.

These art treasures were collected in the eighteenth century during the reign of the Manchu Ching dynasty emperor, Chienlung. There are bronzes and jades of the Shang period which flourished in China more than ten centuries before the birth of Christ. There are also paintings and calligraphy which are more than two thousand years old, and countless other treasures such as porcelain, lacquerware, rare books and documents. It is said that the museum displays can be changed every six months, for thirty years, without a single item being repeated!

The countryside is never too far away from the crowded, bustling city of Taipei. Just over seventeen miles (about twenty-eight kilometers) south of the city is the mountain resort of Wulai, home of the aborigines of the Atayal tribe.

The Atayals are the second largest tribe in Taiwan. They were formerly hunters. Deer, wild boar and bears were some of the animals they hunted in the forests of their mountainous region. Because of their closeness to Taipei, the Atayals of today earn their living from the lucrative tourist trade. They make and sell handicrafts, and perform their ancient tribal songs and dances for tourists.

About thirty-one miles (fifty-one kilometers) southwest of Taipei is Lake Benevolence with its camphor forests and bamboo groves. In this area, too, is the Shihmen ("Stone Gate") Reservoir. Its dam was built with American aid.

Further south is Safari Park, one of the finest wildlife parks in Asia. It has a stock of wild animals not only from Asia, but also from Africa and from North and South America.

Some distance south of Safari Park, lies Lion's Head Mountain. It is important as a Buddhist center. Many of the temples in the area are built into the caves in the mountainside. As its name suggests, its peak, viewed from a certain angle, resembles a lion's head.

To the north of Taipei is the Tatun volcanic range with its hot sulphur springs and fumaroles (smoke-holes found in volcanic regions). The best way to see this region is to take the road north of Taipei.

This road soon divides into two. The right fork winds up the Yangming Mountain past the villas of the rich and a farm famous for its orchids. The air becomes cooler and the mountain scenery more beautiful. At the peak is the Yangming National Park with its cherry trees and azaleas and well-kept gardens. The road continues through forested areas, terraced paddy fields, vegetable plots and quiet villages to the town of Chinshan on the northeast coast.

The left fork follows the Tanshui River northwards to the ancient town of Tanshui at the mouth of the river. Tanshui has, through the centuries, witnessed the presence of the Spaniards, Dutch, British, French and later the Japanese. Today, its biggest attraction is the fine seafood which earns a good living for its fishermen and restaurant-owners.

En route to Tanshui, a visit can be made to Peitou at the foot of

Yangming National Park with its cherry and azalea trees.

"The Bells," sandstone rock formations at Yehliu.

the Yangming Mountain. This hill resort is famous for its hot springs and open sulphur pits.

The coastal road from Tanshui continues northwards past small farming towns and fishing villages. Then, turning in a southeasterly direction, it goes along the coast to Chinshan, Yehliu and Keelung.

Chinshan, where the mountain and coastal roads meet, has one of the finest beaches in Taiwan. Yehliu is famous for its rock formations caused by weathering and erosion, while the port of Keelung is Taiwan's northernmost city. It was known as Santissima Trinidad when it was occupied by the Spaniards from 1626 to 1642.

Today, Keelung is Taiwan's second largest port. It has a population of just over 400,000 whose main occupations are connected with the port trade and its related industries.

Keelung is one of the wettest cities in the world. It has an average of 214 rainy days a year, with much of the rain falling from October to March.

From Keelung, it is possible to take the expressway back to Taipei or to continue, via the Northeast Coastal Highway that runs eastward along the East China Sea and then southward along the Pacific Ocean, to the port of Suao on the island's east coast. Several branches of the road cut westward across the Central Mountain Range to wind their way back to Taipei.

Keelung, Taiwan's northernmost city and second largest port with a population of just over 400,000.

Many minerals such as coal, natural gas, petroleum, sulphur, asbestos and phosphorus have been found in Taiwan. Unfortunately, the quantities mined are not large. Of these, coal is the most important. It is mined mainly near Taipei and its port of Keelung. Other minerals found in northern Taiwan include deposits of copper and gold and a good quantity of sulphur.

Taiwan's other natural resources include forest and agricultural products such as timber, camphor, rice, cane sugar and tea.

5

Central Taiwan

South of Taipei are the green, fertile farmlands of central Taiwan. Agriculture has always been the traditional occupation of the people of Taiwan. However, only one quarter of the land—mainly on the western coastal plains—can be used for farming; the rest consists of forested mountains.

In 1945, when Taiwan was freed from Japanese rule, most of the farms were owned by landlords who rented their land out to tenant farmers. These tenant farmers, who were often poor and lived in mud huts, had to turn over half of their crops to the landlords.

New land reforms were introduced after 1949 by the Nationalist government in Taipei. Farm rents were lowered. In addition, landlords were allowed to own only a certain amount of land. The rest had to be sold to the government who, in turn, resold it to tenant farmers. These farmers were allowed to make their payments in easy installments.

With a fairer distribution of land and improved methods of

cultivation, it soon became possible to have two harvests a year. In later years, farm production greatly increased when large agricultural enterprises were formed and more modern methods of farming were introduced.

Rice, the staple food of the people, is the most important crop grown in Taiwan. It is cultivated in paddy fields covering more than half of Taiwan's farmlands. The twice-yearly yields are enough to feed the population with even some left over for export. Other crops grown are sweet potatoes, wheat, peanuts, soya beans and maize.

Sugarcane is another important crop and so is tea. Tea, completely unknown to the aborigines and early Dutch settlers, was introduced into Taiwan by Chinese immigrants in the seventeenth century.

Tea is China's national drink. The Chinese have been drinking tea for over a thousand years—so much so that it has become very much a part of their culture.

The Chinese drink their tea without sugar or milk. This is very different from the way Westerners enjoy their tea, introduced into Europe during the seventeenth century.

Taiwan produces three types of tea leaves—green tea, *oolong* ("Black Dragon") tea for which Taiwan is famous, and black tea, more commonly known as English tea, which is often drunk in the West. Tea bushes are grown on well-drained mountain slopes in the northwestern hills of Taiwan and on the slopes of the Central Mountain Range.

Hemp and jute are also grown and so are fruits such as

Planting out rice. Rice is the staple food of the Taiwanese and by far the most important crop grown in the country.

pineapples, bananas, peaches, oranges, lychees, longans and watermelons. A wide variety of fruits are canned for export. Vegetables are grown mainly for home consumption.

Today, Taiwan is a major world supplier of canned pineapple, asparagus, mushrooms and many other Chinese food items. Asparagus is grown in sandy ground along seashores and rivers. Mushrooms are cultivated in tiers of earth-filled trays kept in dark, humid sheds.

Central Taiwan's main city is Taichung. *Chung* means "central."

Taichung is the third largest city in Taiwan. It has a population of about 800,000 and is located on the plain about eighty miles (one hundred and thirty kilometers) south of Taipei and ten miles (seventeen kilometers) from the coast. It was founded in 1721 by Chinese immigrants from the mainland.

To the northwest of the city is Taichung Harbour. Completed in 1976, it has opened up central Taiwan to international export markets. Branches of the expressway connect Taichung with its coastal harbor.

About half-an-hour's drive southwest of Taichung is the country town of Changhua, often visited for its famous huge concrete image of the seated Buddha. In the grounds surrounding the image are statues, pavilions, shrine halls and

Tea bushes on the slopes of Taiwan's Central Mountain Range.

A lychee tree. Taiwan's mild climate makes it possible to grow a wide variety of fruits, many of which are canned for export.

pagodas containing paintings and exhibits tracing the history of Buddhism.

To the southwest of Changhua is the ancient port of Lukang ("Deer Port"). It was a popular port of entry for immigrants from the province of Fukien during the Manchu Ching dynasty. It is no longer in use as a port, due to the silting up of its harbor after the Japanese closed it down in 1895. But Lukang retains its old-world charm with its ancient temples and narrow residential lanes.

The people of Lukang are mostly fishermen. Some are artisans who can still be seen fashioning altar tables, shelves, ornaments and furniture, using ancient tools and techniques. Others make incense for use in the temples of Taiwan.

About fourteen miles (twenty-two kilometers) south of Taichung is Chung Hsing Village, the seat of Taiwan's provincial or local government.

Central Taiwan's best known holiday resort is Jih Yueh Tan ("Sun Moon Lake") situated in the western foothills of the Central Mountain Range. About 2,500 feet (762 meters) above sea level, it is in the county of Nantou, one of the sixteen counties into which Taiwan is divided. Seen from the surrounding hills, the lake is either round like the sun or crescent-shaped like the moon.

There were originally two lakes, until the Japanese constructed a dam in the early twentieth century for hydroelectric purposes. This raised the water level and merged

The giant statue of the Buddha at Changhua.

the two lakes into one. The area is a good base for hiking, as well as for visits to an aboriginal settlement and to Taiwan's highest pagoda and biggest temple.

Also in central Taiwan is the East-West Cross-Island Highway. This is Taiwan's engineering marvel, cutting through the Central Mountain Range to link the eastern and western coasts. This 120-mile (193-kilometer) long highway took ten thousand workers four years to complete. The Taiwanese call this highway the "Rainbow of Treasure Island." Why "Treasure Island"? Because that was what the early settlers called Taiwan—*Bao Dao*, their "Treasure Island." And the rainbow is,

The Tzu En Pagoda at Sun Moon Lake. *Tzu En* means "filial devotion," which the Chinese regard as a supreme virtue.

of course, made up of the spectacular colors of Taiwan's green mountain valleys and alpine forests, its rocky ravines and gushing river torrents, its lakes and snow-capped mountain peaks, and its villages and towns along the route.

The highway starts from Taroko on the east coast. It cuts across the mountains to the town of Tungshih, northeast of Taichung. Lishan ("Pear Mountain"), some 6,380 feet (1,945 meters) above sea level, is the midway station of the highway. Here, the highway divides into two, one branch going northeast to the country town of Ilan and the Northeast Coastal Highway, and the other in a southwesterly direction towards the western plains.

Lishan is a popular mountain resort. To the south is the village of Tayuling. The highway from Tayuling goes south around Hohuanshan ("Mountain of Harmonious Happiness"). Looming 11,224 feet (3,420 meters) above sea level, it is Taiwan's only winter ski resort where, during the two short months of January and February, skiing is possible because of the heavy snowfall.

From here, the highway cuts in a southwesterly direction towards Puli, a mountain village known as the exact geographical center of Taiwan. After Puli, the highway continues westward to the coastal plain, while a branch goes southward to Sun Moon Lake and then westward, through the tea-producing region of Sungpoling, to join the coastal road. The best Taiwanese *oolong* tea is said to come from the Sungpoling region.

Hohuanshan, the Mountain of Harmonious Happiness. This is Taiwan's only winter ski resort with heavy snow in January and February.

Forests, which cover about fifty-five percent of Taiwan's mountainous area, provide the country with valuable timber, and with bamboo and camphor products.

Bamboo and trees of camphor laurel, sandalwood, oak, teak and black ebony grow in the lower reaches of the mountains. Higher up, where the climate is cooler, forests of cedar, cypress, pine and conifers become more common.

In the mountains of central Taiwan are three experimental forest areas. The most important is the Chitou Bamboo Forest which supplies almost forty percent of Taiwan's bamboo and bamboo products. At 3,774 feet (1,150 meters) above sea level, this area lies between Taichung and Chiayi—a city in the southern part of central Taiwan. The research station is run by the National Taiwan University. Many varieties of bamboo are cultivated in this region, along with trees such as the cedar, cypress, pine and other evergreens. Every year, more than a million tree shoots are distributed for Taiwan's reforestation projects.

The bamboo, a giant tree grass, is a very useful plant. Its young, tender shoots can be eaten. Its long, hollow stems can be made into paper and a thousand and one other things such as chopsticks, handbags, ornaments, fans, baskets, trays, furniture, boats and bridges. Strips of bamboo can also be made into mats and ropes.

Visitors to the village of Chitou can find in its shops bamboo goods and other products including tea, mushrooms and herbs. They can also go hiking or camping, and enjoy the cool mountain air and scenic beauty of the area.

The second experimental forest area is in the highlands north of Puli. The third is in the Alishan (Mount Ali) forest of cedar, cypress and pine in the southern part of central Taiwan.

The Alishan area can be reached either by road or by a narrow-gauge railway from the small, manufacturing city of Chiayi. The more interesting route would be by train, which takes about three hours. The train crosses eighty bridges and passes through fifty tunnels, one of which is 840 yards (768 meters) long.

The railway station at Alishan is 7,187 feet (2,190 meters) above sea level, making it the highest railway station in the Far East. About twenty-five miles (forty kilometers) to the east of

Alishan is Yushan ("Jade Mountain"), also known as Mount Morrison. It is the highest mountain in Taiwan and is a favorite with mountain-climbers.

Wu Feng, an interesting historical figure, known as the Loyal Lord of Mount Ali, is worshipped at a Chinese shrine which can be seen along the highway south of Chiayi, on the way to Alishan. He was born in 1699 in the province of Fukien in China. He came to Taiwan as a youth and later became the interpreter

A bamboo forest. Many varieties of bamboo are cultivated in Taiwan.

Dense forest of cedar, cypress and pine in the Alishan region. Forest covers about fifty-five percent of Taiwan's mountainous area.

and liaison officer of the Chinese settlers on the plains and the aboriginal tribes in the mountains. He worked hard to stop hostility between the two.

At the age of seventy-one, Wu Feng thought of a plan to stop, once and for all, the aboriginal practice of invading the plains to hunt for Chinese heads as sacrifices to their gods. Wu Feng told his aboriginal friends that they would see a man "wearing a red hood and cape, and riding a white horse," on a certain day and at a certain time and place. They were to cut off the man's head. The aborigines did so, only to find that their victim was none other than their old friend, Wu Feng. His courage and self-sacrifice persuaded the aboriginal tribes of the Alishan region to give up their practice of head-hunting.

Wu Feng is believed to be the only historical figure worshipped by both the Chinese and the aborigines of Taiwan. His birthday is celebrated every year on November 12.

In this region of central Taiwan, in a low mountain pass between Chiayi and Shan Hu Tan ("Coral Lake"), is the hot spring spa of Kuantzuling. Within three miles (five kilometers) of the spa is the Chi Yen Yi Shih ("Exotic Rock"), a boulder as huge as a house. On its sides can be seen the preserved remains of prehistoric plants and animals. Not far from this rock is the Shui Huo Tung Yuan ("Water-Fire Crevice") with its boiling mineral water and constant flickering flame. Both are reminders of Taiwan's volcanic past.

One last interesting place to visit before leaving central Taiwan is Peikang ("Northern Port"). It is to the northwest of Chiayi. In Peikang is a very well-known temple dedicated to Matsu, Goddess of the Sea. It is believed to be the richest temple in Taiwan, visited by more than three million pilgrims every year.

Religious rites, unchanged since early times, are performed at the Peikang temple throughout the year. The most important festival is the birthday of the goddess Matsu. It is celebrated throughout the three hundred and eighty-three Matsu temples in Taiwan in April or May, with the biggest celebrations taking place in Peikang.

6

Southern Taiwan

Southern Taiwan is the earliest settled region of the island. Its most fascinating city is, of course, Tainan–the oldest city in Taiwan. It was the island's ancient capital from 1663 to 1885.

Not far from Tainan, to the northwest, is Anping the site of Fort Zeelandia, built by the Dutch in 1624. The fort has been completely restored. Its only original relic is the remains of an ancient wall. Today, Fort Zeelandia serves as a memorial to Koxinga.

Koxinga, the Ming loyalist, retreated to Taiwan in 1661 with thirty thousand troops and eight hundred war junks (sailing vessels). He drove the Dutch out and set up his court and government in Tainan in defiance of the Manchus who had conquered China. Tainan grew into an important center with Anping as its port. Due to the silting up of its harbor, Anping, like Lukang and Peikang, is no longer in use as a port today.

Tainan is Taiwan's fourth largest city. It has a population of about 700,000. Steeped in culture and history, it has become a popular tourist destination. Although known as the "City of a Hundred Temples," Tainan has many more temples than that. There are two hundred and nine major temples and many minor shrines in the city and surrounding countryside. The oldest Confucian temple in Taiwan can be found here. It was built in 1665 and is believed to be the best example of Confucian temple architecture.

The Koxinga family ruled Taiwan until 1684 when the Manchus from the mainland finally conquered the island. Almost two hundred years later, in 1875, the imperial Manchu court allowed a shrine to be built for Koxinga. The former Ming resistance leader had been forgiven and could now be recognized as a national hero. Koxinga's shrine is not far from the temple of Confucius.

There are many other ancient temples and monasteries in Tainan, some more than three hundred years old. They are preserved by a careful program of renovation work. As it is the island's most traditional city, Tainan's religious festivals are more frequently and elaborately observed than are the same festivals in the north.

Further inland, to the northeast of Tainan, are Coral Lake and Tsengwen Reservoir—two of the largest lakes in Taiwan. This area, created by man and nature, is one of the loveliest regions on the island.

From the air, Coral Lake looks like a coral formation, hence its name. It receives its main water supply from the nearby Tsengwen Reservoir built in the foothills of the Central

A street scene in Kaohsiung, southern Taiwan's main city.

Mountain Range. The water is distributed to the surrounding area for use in homes, agriculture and industry.

To the south of Tainan is Kaohsiung, the main city of southern Taiwan. It is the island's largest international port, and the second largest city after Taipei. It has a population of about 1.5 million and, like Taipei, is a "special municipality" which makes it equal in status to a province. Taipei was made a special municipality in 1967 while Kaohsiung became one in 1979. Kaohsiung is a more modern city than Tainan. It has broader streets, more open spaces and buildings of steel and glass. It is the only city in Taiwan, besides Taipei, with an

A general view of Kaohsiung.

international airport. It is also the terminus of the north-south expressway.

Kaohsiung grew rapidly because of its industries. Aluminum processing plants, oil and sugar refineries, shipbuilding yards and steel mills are among its major industries.

To cut down on pollution problems, Kaohsiung is now attracting high technology industries to its central districts while moving its heavy industries to new industrial zones in the suburbs. A sandbar in Kaohsiung's harbor has been developed into an industrial area producing goods for export. Many other such zones have been built all over Taiwan.

Kaohsiung is also the world's largest scrapper of old ships. Steel, copper wires, screws and other parts are recovered and sold as scrap metal. Nothing useful is ever wasted. Kaohsiung

78

also has a dry dock which is the second largest in the world and a container port which is the fifth largest.

The Japanese started Taiwan on the road to industrialization. With the encouragement of the Nationalist government, industrial growth continued, especially in the areas surrounding the harbors of Keelung and Kaohsiung–Taiwan's two main ports.

By the 1970s, Taiwan had shifted from a traditionally agricultural economy to an industrial one. The fastest-growing manufacturing industries were textiles, chemicals, canning and wood-processing. Today, Taiwan's attention is directed towards

A woman at work in an electronics factory in Taipei. Although agricultural exports are still important, manufactured goods now make up more than fifty percent of Taiwan's export trade.

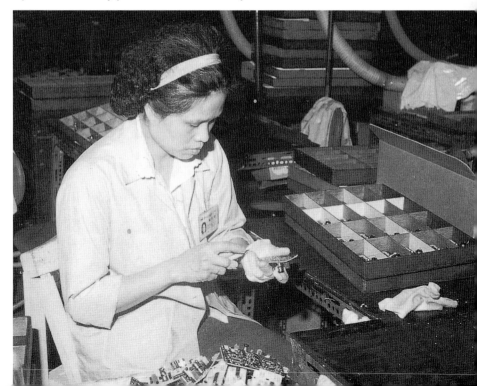

such industries as computers, electronics, precision machinery, communication equipment and automobiles.

Taiwan's best business has, since the early days, been trade. Major exports have included sugar, coal, rice, tea, fruits, tobacco, timber and camphor. Although agricultural exports are still important, manufactured goods now make up more than fifty percent of Taiwan's foreign trade.

Taiwan's major trading partners are the United States, Japan, the countries of the Association of Southeast Asian Nations (Thailand, Malaysia, Singapore, Indonesia, Brunei, the Philippines) and Europe.

Besides being Taiwan's largest industrial center, Kaohsiung is also an important fishing port. Many fishing vessels leave Kaohsiung for rich fishing grounds in the seas near by. Some go as far away as South African waters.

Like agriculture, fishing has been a traditional occupation of the people of Taiwan. Many fishing villages and towns can be found along Taiwan's coastline. Fish are also caught in the island's lakes and rivers and fish-farming has greatly boosted Taiwan's fishery production. Taiwan's catch each year easily totals up to a million tons and is sold to both local and overseas markets.

Southern Taiwan enjoys a sunny, tropical climate. It is hotter and more humid than the northern region. Much of the rain falls between May and September, brought by monsoon winds or typhoons that blow during this period.

Baby sharks caught off the Hengchun Peninsula.

Like central and northern Taiwan, southern Taiwan has a road that cuts across the island to link the southwestern coast to the southeastern coast. It is called the Southern Cross-Island Highway. It was completed in 1972 and makes travelling from Tainan or Kaohsiung to Taitung on the east coast much easier and faster.

The Central Mountain Range tapers off into the Hengchun ("Eternal Spring") Peninsula in the south. Fishing villages, sandy and rocky beaches, forested hills and coral rock formations dominate the scenery in this narrow southern region.

81

Here too, is the Kenting National Park. It was started by the Japanese in 1906. It has some remarkable coral formations, one of which is known as the "Fairy Cave." It was carved out by the action of waves, when the area lay deep beneath the sea thousands of years ago.

The Park also has a large collection of rare plants from all parts of the world. The coastal road south of Kaohsiung takes the visitor down south to Kenting Park and from there to Oluanpi ("Goose Bell Beak"), the southernmost tip of the island of Taiwan.

7

Eastern Taiwan

Eastern Taiwan has often been described as "aborigine country." Its mountains and inland valleys are home to many aboriginal tribes, the largest of which is the Ami.

Mountains cover about ninety percent of the region. There are narrow strips of flat land on the coastal plain, separating the mountains from the sea. The longest strip stretches between the towns of Hualien and Taitung.

Cut off from the rest of Taiwan by the great mountains of the Central Range, this rugged eastern region has been less affected by modern development than the other parts of the island. Lifestyles remain different. So do the art, language and culture of the people.

In the northernmost part of eastern Taiwan is the harbor town of Suao with its fisherman's wharf where all kinds of seafood are available. The town forms a major link on the railway line from Taipei to Hualien and further south to Taitung. This small harbor town is the fifth and newest port

Fishing boats in Suao harbor.

in Taiwan after Kaohsiung, Keelung, Taichung and Hualien.

Hualien, the largest town in eastern Taiwan, is about 50 miles (80 kilometers) south of Suao. Connecting the two towns is the East Coast Highway. First built in 1920, the road winds along the steep cliffs of the eastern coastline some 1,000 to 1,500 feet (300 to 450 meters) above sea level. Traffic travels along this narrow cliff road in one direction only, either northward or southward, at certain times of the day. The road is being widened to enable traffic to move in both directions along parallel lanes at the same time.

About nine miles (fifteen kilometers) before reaching Hualien, a branch of the highway cuts westward into the gorge of Taroko. This is the start of the East-West Cross-Island Highway. The highway was built not only to connect the east coast with the west coast but also to open up the great mountainous area to farmers, lumberjacks and tourists.

The stretch through the Taroko Gorge is the most popular section of the highway. The gorge, through which the boulder-strewn Liwu ("Foggy") River gushes, winds for twelve miles (nineteen kilometers) from the coast to its upper end at Tienhsiang. Beyond this spot, at the bottom of the gorge, are the hot springs of the area.

The entrance to Taroko Gorge.

The East-West Highway cuts and tunnels its way through the towering marble cliffs that line both sides of the spectacular gorge. There are thirty-eight tunnels on this stretch, some with windows carved out of the mountain itself so as to provide light and ventilation. The mountain scenery is breathtaking. Taroko Gorge seems to be just the right name for it—*Taroko*, in the Ami dialect, means "beautiful"!

The coastal town of Hualien occupies the narrow strip of flat

Part of the East-West Highway, which cuts and tunnels its way through the towering marble cliffs that line the Taroko Gorge.

A marble tile factory in the coastal town of Hualien—the marble is quarried from the cliffs of nearby Taroko Gorge.

land along the coast. It is famous for its marble, quarried from the cliffs nearby Taroko Gorge. Hualien produces marble goods such as lamps, bookends, ashtrays, vases and coffee tables. Many of its hotels and temples are built of marble, and so is its domestic airport! Tourists en route to Taroko Gorge usually fly to Hualien from either Taipei or Kaohsiung.

Hualien is particularly festive in late July and early August when the Ami celebrate their harvest. The Ami are mostly farmers, growing crops such as millet, sweet potatoes and mushrooms; and fruits such as pears, peaches and plums. During their harvest festival, visitors can see them dressed in their bright, colorful ceremonial costumes and headgear, and watch them celebrate their festival with tribal dances.

An Ami woman using a traditional loom to weave colorful material.

Other smaller coastal towns and villages where the Ami live lie to the south of Hualien. Shops in these towns offer an interesting choice of their aboriginal art and craft.

South of Hualien, the highway divides into two—one branch follows the coastal route and the other the foothill route about nineteen miles (thirty kilometers) inland. Both routes converge onto the small, seaside town of Taitung.

The hillsides and lowlands between Hualien and Taitung are well-cultivated. Farmers plant rice, sugarcane, vegetables and fruits on every available plot of land. Fishermen in the villages along the east coast set out in their boats for the fishing-grounds of the Pacific Ocean. The seas of eastern Taiwan are rougher

88

than in other regions and the weather less predictable, but fish is plentiful.

Taitung itself is a small, quiet town. It is a convenient stopover for people visiting the islands off its coast or en route to enjoy the hot springs or to explore the mountain peaks and valleys to the south.

Although eastern Taiwan is the largest region on the island, it is the least populated. Mountains and foothills occupy most of the region which still has many remote, unspoiled places. Wildlife, similar to that of the southern Chinese mainland, includes deer, bears, wildcats and panthers. There are also pheasants, kingfishers, larks and birds of many other species.

8

Island Outposts

Taiwan's largest group of offshore islands is found off the midwestern coast, in the Taiwan (Formosa) Strait between the Chinese mainland and the island of Taiwan. This archipelago consisting of sixty-four islands was called the *Pescadores* ("Fishermen Isles") by the Portuguese. Its official Chinese name is the Penghu Islands.

The islands, twenty-one of which are uninhabited, have a total land area of 49 square miles (127 square kilometers). They have a population of about 120,000, more than half of whom live on the largest island of Penghu, mainly at Makung, the only big town in the group of islands.

Most of the islands are flat and rocky. There is not much land suitable for agriculture. There are some small vegetable plots, but they have to be well-protected from the winds that sweep the islands. These winds are especially strong during the months of October to March. They blow from the northeast at speeds that can reach as high as over 40 miles (70 kilometers) an hour.

Typhoons often hit the islands during the summer months of June to October.

The Penghu Islands became Taiwan's sixteenth and smallest county in 1960. It is the republic's only island county. Makung, its capital, has a naval base and an airport. A bridge links the main island of Penghu to the next two large islands—Paisha and Hsiyu. This bridge is known as the Penghu Bay Bridge. It is the longest inter-island bridge in the Far East.

Many of the early settlers who came to Taiwan used the Penghu Islands as a stepping-stone. The first Hakkas stopped here and so did the Dutch, Koxinga and his Ming loyalists, the Manchu forces, and the French. Many of their temples and monuments can be found in various parts of the flat, barren islands.

The oldest temple was built in 1593 to honor the goddess Matsu. There is also a temple to Confucius, built in 1767, and one to Kuan Yin, the Goddess of Mercy, built in 1885.

Fishing is the main occupation of the islanders. Rich fishing grounds lie to the north and northwest. As a result, fresh seafood is always available at the small restaurants on the islands.

The warm, shallow waters around the islands favor the growth of coral. Coral is a hard substance secreted by tiny sea animals around the outside of their soft bodies. Billions of these tiny animals live together in clumps called colonies. When they die, their skeletons remain to form corals of many sizes, shapes and colors.

This coral is collected by the fishermen. Master craftsmen

fashion it into rings, pendants, necklaces, earrings, bracelets and a wide variety of carved figures. The Chinese consider pink coral the best, with red coral coming a close second. Besides being pretty, coral is believed to bring good luck to its owner.

Tourism has greatly helped in improving the standard of living of the islanders. There are now regular air services linking Taiwan to Makung. There are also regular ferry services between Kaohsiung and Makung.

Some of Taiwan's other offshore islands open to visitors are Little Liuchiu, Green Island and Lanyu or Orchid Island. Little Liuchiu is a wooded islet about twenty-two miles (thirty-five kilometers) due south of Kaohsiung. Green Island and Orchid Island are off the southeast coast of Taiwan.

Green Island is within sight of Taitung, to the east. It has been recently developed for tourism and can be reached by sea or air from Taitung. It was originally known as Fire Island because beacons had to be lit on the island to warn fishing vessels about its coral reefs.

The island is very small but its hills are popular with hikers and its waters and reefs are ideal for swimming, fishing and scuba diving.

Lanyu, or Orchid Island, named after the wild orchids that grow there, is situated southeast of Taitung. It is mostly hilly and is the home of the Yami, the smallest and most primitive of Taiwan's aboriginal tribes.

The Yami have not been as greatly influenced by modern civilization as the other aborigines because the Japanese isolated

Part of Green Island off the southeast coast of Taiwan—the black rocks are evidence of the island's volcanic origins.

Lanyu during their occupation of Taiwan in the early twentieth century. The Yami were left to live in the simple manner of their ancestors who are believed to have come from Polynesia, the Pacific islands east of Australia.

There are about 2,600 members of the Yami tribe on Orchid Island. Although the government has built housing blocks for them, they prefer their traditional homes built underground into the hillsides. These homes are well-protected against the typhoons that rage across the island every summer.

The Yami women grow taro, which resembles the sweet

potato, as well as other vegetables and fruits. Their men go fishing in boats which they carve out of tree trunks, using ancient hand tools. These fishing boats are their most important possessions. A boy becomes a man when he completes his first boat.

It is only in recent years that the Yami have begun to use the tools of modern man. In addition, the Yami men now wear shorts or trousers instead of their traditional loincloth.

Because of military installations on certain parts of the shoreline, some areas of Orchid Island are out of bounds to visitors as are a few other offshore islands of Taiwan. These include the Kinmen (Quemoy) and Matsu Islands.

The Kinmen Islands are a group of twelve islets blocking the mouth of Amoy Bay. These islands are very close to the Chinese mainland. The main island of Kinmen is shaped like a dumbbell. It has both arable land and boulder-strewn areas. Mount Taiwu in the center rises to about 830 feet (253 meters). The other islands are low and flat.

The first inhabitants of the Kinmen Islands came from the Chinese mainland during the fourth century. Today, their descendants speak the Amoy dialect. They are mostly farmers growing crops such as sweet potatoes, peanuts, barley, wheat, soya beans and vegetables. The islands are self-sufficient in all foods except rice.

These islands served as Koxinga's base in his fight against the invading Manchu forces during the seventeenth century. In the mid-twentieth century, the Communists made several attempts to

Fishing boats of the Yami who live on Orchid Island.

take the Kinmen Islands from the Nationalists who were holding them. They never succeeded.

The Matsu Islands are further north, not far from the mouth of Communist China's Min River. They consist of nineteen hilly islets. The main islet is Nankan, more commonly known as Matsu. The main occupation of the islanders is fishing.

Taiwan's garrison troops on these island outposts keep a watchful eye on Communist China. The soldiers are well-trained and fully equipped to defend Taiwan's democratic way of life.

9

Taiwan's Problems

Different ideas about how China should be ruled have, in recent history, created a huge rift among the Chinese people. Today, the Nationalists rule Taiwan and the Communists rule mainland China. Both agree that Taiwan is a province of China. Both wish to see Taiwan united with mainland China, but only under their own political systems.

The Nationalist government of Taiwan (the Republic of China) and the Communist government of mainland China (the People's Republic of China) both claim to be the only rightful government of the whole of China. Taiwan insists that before formal talks can be reopened between them, Communist China must abandon communism and accept Dr. Sun Yat-sen's "Three Principles of the People" as the foundation of a modern Chinese state.

This is unacceptable to Communist China which has never, in public, renounced the use of force against Taiwan. Threatened politically and militarily by the Communist regime on the mainland, Taiwan has built up a military force dedicated not only to the defence of the island republic but also to the eventual recovery of the mainland. Its military expenditure has given Taiwan one of the heaviest defence burdens in the world.

The United States, Taiwan's powerful ally, has given the Nationalist government in Taipei both military and economic aid. In the early years, it not only helped to protect Taiwan but also led a crusade to keep Communist China out of the United Nations Organization.

But in 1971, more countries, hoping to improve their trade, voted to recognize Communist China as the official China. As a result, Taiwan was expelled from the United Nations. In 1979, the United States also finally decided to recognize Communist China. It then severed all diplomatic ties with Taiwan.

Taiwan looks on the United States as its trusted friend and ally against communism. Recent moves to further strengthen relations between Communist China and the United States have made Taiwan very unhappy. But it continues to stand firm against Communist China.

Although the United States has ended its mutual defence treaty with Taiwan, it has kept to its other agreements. It continues to sell Taiwan advanced weapons for the island's defence and is increasing its "unofficial" contacts with the republic. Many other countries have also opened trade offices in Taipei, although their governments officially recognize only Communist China.

Despite its political problems, Taiwan's democratic system of government has led to the island's booming economy. Taiwan has become a self-supporting country. It no longer depends on foreign aid. Instead, it gives technical assistance to developing countries in Asia and Africa.

Today, the Taiwanese enjoy one of the highest standards of living in Asia. Like the islands of Singapore, Hong Kong and Japan, Taiwan has become one of Asia's giants in the field of economic progress.

GLOSSARY

aborigines   The first or earliest known inhabitants of a region.

alluvial soil   Soil deposited by running water on an area of land.

Analects   The collection of the teachings and sayings of Confucius.

Confucius   Considered to be China's greatest teacher, he lived from 551-479 B.C. His teachings are still followed in Taiwan today.

Formosa   Name given to an island by 16th century Portuguese sailors which would later become known as Taiwan.

hydroelectric power   Electricity generated by water power.

joss stick   A slender stick of incense.

junk   Sailing vessel, boat.

maize   Corn.

monsoons   Seasonal wind storms.

taro   A vegetable resembling the sweet potato.

totems   An object that is an emblem or revered symbol.

INDEX

100

tea, 48, 61, 63, 69, 71, 80
Teachers Day, 46
temperatures, 27, 28
temples, 43, 44, 46, 57, 66, 68, 74, 76, 87, 91
Teng-hui, Lee, 11, 12, 22-23
Three Principles of the People, 22, 30, 96
Tienhsiang, 85
tourism, 57, 76, 87, 92
trade, 18, 41
tribes, 33, 34
Tropic of Cancer, 26
Tsengwen Reservoir, 76
Tungshih, 69
typhoons, 29-30, 80, 91, 93

U

United Nations Organization, 11, 97
United States, 8, 10, 11, 12, 18, 21, 22, 80, 97

V

vegetables, 64
volcanic activity, 13, 74

W

warlords, 19
weaving, 34, 88
West, Westerners, 15, 17, 18, 40, 41, 63
wildlife, 57, 89
Wu Feng, 72, 73, 74
Wulai, 34, 57

Y

Yami tribe, 34, 92, 93
Yangming, 57, 59
Yehliu, 25, 59
yuan (councils), 8, 30, 31
Yushan (Jade Mountains, Mount Morrison), 26, 27, 72